THE STATE OF NEVADA

PROOF OF APPROPRIATION OF WATER FOR IRRIGATION

From ...... *Meadow Valley Creek* ......
Name of natural water source

Through ...... *Ditch and Flume* ......
Name of ditch, flume, or pipe line

...... *I L Carson* ......, the undersigned, being first

duly sworn, deposes and says that the facts relative to the appropriation of water by ...... *He mi* ......

...... are full and correct to the best of his

knowledge and belief.

If proof is not made by claimant, deponent should state on this line by virtue of what authority he represents the claimant

(1) Name of claimant ...... *Leonard L. Carson* ......

Address ...... *Caliente* ......, County of ...... *Lincoln* ......

State of ...... *Nevada* ......

(2) The means of diversion employed ...... *Ditch and Flume* ......
Dam and ditch, pipe line, flume, etc.

(3) The date of the survey of ditch, canal, or pipe line was ...... *Not known* ......

(4) The construction of the ditch or other works was begun ...... *Over 49 Years ago* ......

and completed ...... *Over 49 Years ago* ......

(5) The dimensions of the ditch or canal as originally constructed were: Width on bottom ...... *2 1/2 feet*

feet, width on top ...... *4* ...... feet, depth *2 1/2* feet, on a grade of ...... *10* ...... feet per thousand feet.

(6) The conduit has (has not) been enlarged.
NOTE—If enlargement or extension of ditch was made, supply information under (7) and (8)

(7) The work of enlargement of the ditch or canal was begun ......

and completed ......

(8) The dimensions of the ditch or canal as enlarged are: Width on bottom ...... feet, width on

top ...... feet, depth ...... feet, on a grade of ...... feet per thousand feet.

(9) The claimant is (is not) an owner in the above-described conduit.

...... *I am sole owner* ......
If claimant is an owner in the conduit, state interest held on this line

(10) The nature of the title to the land for which the water right is claimed is ...... *Possessory Claim* ......

(11) Crops of ...... *Luceme Corn Vegetable and Fruit of all kinds*

have been grown upon the land irrigated.

(12) The water has been used for irrigation from ...... *about March 1st* to ...... *October 1st*
Day of month   Day of month

of each year.

(13) The water was first used for irrigation by claimant or ...... grantors in the year *1898*

when ...... *35 "* ...... acres were irrigated in the ...... of Sec ......,

T ...... R ...... E. ...... *The ground has never been*

HOW TO BUY WATER RIGHTS

Chris W. Miller

Millers Rock Farm, LLC

Published by Millers Rock Farm, LLC

ISBN 13: 978-0-9970949-0-9

Please visit www.westernstateswaterrights.com or contact Chris Miller at chris89027@gmail.com for more information.

Book design and production assistance by Adam Robinson for Good Book Developers.

To Dad

He started me out in life with a love, respect, and deep appreciation of nature.

"Whiskey's for Drinkin' and Water's for Fightin'"
—*Author Unknown*

"You are piling up a heritage of conflict and litigation over the water rights. There is no sufficient water to supply the land."
—*John Wesley Powell to the International Irrigation Congress in Los Angeles in 1893*

"When the Well's Dry, We Know the Worth of Water"
—*Benjamin Franklin*

CONTENTS

PREFACE

THIS IS INTENDED TO BE YOUR GUIDE TO A MORE secure water future. Water is a vital element to sustaining life. Nothing lives without water. It is my attempt to simplify the very convoluted and confusing world of water rights. This book is designed to help you answer water related questions, solve water supply problems, find water opportunities, and avoid serious mistakes. It is written for the average person to present the complicated matter of water rights in a straightforward way.

Only about three percent of the Earth's water is fresh water. The rest is in our oceans. Of this three percent fresh water, approximately two thirds is frozen, and much of the remainder—about one percent—is locked underground. It is not where we need it to be and it is not easy to move around. Drought, over allocation, and over pumping have left huge areas of the world with critical fresh water shortages. Entire municipal water districts serving large metropolitan regions with millions of residents are losing their fresh water supply to falling water tables. For example, Sao Paulo, Brazil—home to twenty million people—is rationing water. In the Southwest United States surface water storage systems like Hoover Dam's Lake

Mead, the largest of our man made reservoirs, tend to give a false sense of security.

And as for our diet, most of us are meat eaters and this demand is increasing around the world, as the middle class expands in places like China and India. Meat consumption is expected to double by 2050 due to the expanding middle class. A human diet which includes meat requires sixty percent more water than a vegetarian diet. To put it simply, this is a much greater hazard to our environment than fossil fuels. Two thirds of our fresh water is used in food production and eighty million hungry people join the world's population every year.

Accordingly, demand for and pressure on this finite precious resource continues to grow. For many, this is about personal responsibility. Are you water dependent to the point that if you turn on the tap and nothing comes out, you are out of water and options? When you own the rights to put water to your beneficial use you have gained the ability to produce food for your family, as well as the feed for your food-producing animals, like chickens which lay eggs or the beef, pork, and chicken you eat. Food production gives you a huge level of personal control. You can raise your own organic plant-based diet. Control of the water gives you choices and the ability to be self-reliant. The fact that you are reading this book indicates you have a level of awareness and concern about being in control of your life. Also, you know our waters are in trouble.

In these pages, I'll share information on what you need to know about water rights so that you can avoid the horror stories that befall the people who never asked, or even wondered, how water, our lifeblood, is allocated. I'll share many of those stories here, and once you have read this book, you will have the tools, the knowledge, and confidence to go out and secure your own water future.

CHAPTER 1

Will There Be Water?

THE SCIENCE AND FACTS ARE CLEAR, FROM THE data provided by the NASA satellite GRACE, which stands for Gravity Recover and Climate Experiment, to the tree ring studies looking back a thousand years: the world's fresh water supply has serious problems. Generally, water is not fairly priced—in many regions it is far too cheap. Many of the world's greatest aquifers, including those under our best farmland, are being pumped dry. Water, along with the food it produces, will inevitably cost more, regardless of who supplies it. As time goes by, demand increases, water tables fall, aquifers fail and wells go dry. Higher fees, restrictions and limitations on you and your water consumption are guaranteed. Over allocated ground and surface water is an extremely widespread reality today.

In many areas we in the United States already see very restricted use. In other areas, water will simply not be available and accessible for consumption of any kind. The EPA is pushing hard to increase the agency's reach by changing the definition of "navigable waters" to take full control of every drop that reaches the surface, whether by precipitation, pump, or natural flow of artesian spring water. New regulations, restraints, restrictions, and fees are making it impossible for many farmers and individuals to keep up the practices of the past. Today, if your land has access to a municipal or private culinary or residential water system, you will not be allowed to drill your own water well on your property, in most cases. In most states the only time you can hire a well driller to drill a water well on your property is conditional on not having access to a public water system.

The question of whether there will be water is not as alarming as it sounds. The problem is that the water is not where we need it to be. In fact, there will be abundant fresh water in many places. At the same time, in the increasing number of aquifers and watersheds where fresh water is severely scarce, there are going to be extreme limitations and restrictions. This is very important in terms of where you decide you want to live. This is about your location selection. With water, just like real estate, location is everything. If the place of use location—a certified surveyor's map of the exact place for specific water to be consumed—has a reliable watershed, (surface water) and aquifer (ground water) and the rights are in good standing, then you have relative water security.

These two systems, surface and ground water, are mapped on two different mapping systems. While both may be referred to as basin maps, one is a surface drainage region, the other is a mapping system of underground water regions. Please make this distinction in your studies. You must understand the amount allocated in the basin compared to the water supply or available water in the basin, the underground, and surface water system to evaluate the balance of your location. Your state charts the underground basin maps and they will provide this information. Drainage watershed or basin maps can be found at the United States Geologic Service (USGS) site. The Columbia River Drainage basin is 258,000 square miles and within this surface drainage area there are dozens of state designated underground basins.

Each state has basin maps. These maps should not be confused with drainage basin or watershed maps. The basins are the mapped out areas, and aquifers are the fresh water below. Basin numbers identify location of both ground and the surface water rights within this mapped area. Aquifer recharge comes from adjoining basin inflows, rain, and snow. The precipitation falling onto the surface drainage basin or watershed can easily take hundreds of years to reach the aquifer. What does not filter down into the aquifer remains on the surface and it is either lost to evaporation or it flows into our rivers, lakes, and streams. As we drill wells into our aquifers we are dropping straws in and drafting the water out. When the shallow underground aquifers are depleted, then the flowing seeps and springs on the surface often dry up.

Today, science tells us that much of the existing groundwater is being depleted due to over allocation and over pumping, resulting in a falling water table.

One of the world's largest aquifers, the Ogallala for example, was charged—or filled—with ice age glacial melt ten thousand years ago. California's Central Valley and the Midwest's Ogallala alone make up over 40 percent of America's irrigated farmland and some of the best producing ground on our planet. Much of the nation's grain and produce come from these two regions. Both are over allocated and both are at risk of being depleted from over pumping. All the headlines and drought news comes down to one thing: there is no sufficient water. Many regions of the world's fresh water supply are at extreme risk. We are consuming water faster and in larger amounts, in the most critical places.

CHAPTER 2

Water Awareness, How Much Water Do You Need?

THE STANDARD MEASUREMENT UNITS FOR YOUR water will most likely be in "acre feet." An AFA or Acre Foot Annually is one acre with one foot of water standing on it. It is 43,560 cubic feet of water, or approximately 325,000 gallons per acre foot. You will need this terminology. An AFA of water amounts to twelve inches of precipitation per year, per acre.

For many people, sustainability is not an environmental or "green movement," it is about themselves and their ability to be as self-sustainable in their lives as possible. Given the recent pace of change, your opportunity to secure your water future may be fleeting.

To determine your water supply needs, you should know your current water consumption numbers and your future projected needs. Very few people know their water footprint. National Geographic's Special Edition of November 1993, and Special Issue of April 2010 were efforts to enlighten the public about their water consumption.

Your Footprint

- To produce a single bed sheet requires 2800 gallons.

- A pair of blue jeans, 2900 gallons.

- One tee shirt, 766 gallons.

- One pound of hamburger, 1857 gallons.

- One glass of milk, 53 gallons.

- One glass of wine, 32 gallons of fresh water.

In Florida 3000 gallons are used to water grass for each golf game played. Dairy cows eat alfalfa hay, and the average alfalfa farmer pumps three to six acre feet of irrigation water to irrigate one acre. In other words, they pump between 975,000 and 1,950,000 gallons of fresh water per growing season, on each acre, to grow hay, to feed cows, so you can drink milk and have cheese on your pizza. This is your virtual water consumption. By the time you get out of bed, put on your jeans and tee shirt, have three meals and a glass of wine, you have consumed thousands of gallons of water—and that does not include showering or flushing a toilet! In fifteen years nearly two billion people around the world will live in regions of severe water scarcity. Will you be one of them?

Your planned use of the water will determine how much water you will need. Future plans of expansion may be a reason to own additional water rights, but be aware of the "use it or lose it" provision for groundwater. To determine your needs, you will need to answer these questions: 1) the amount of land you plan to irrigate; 2) the type of crops and growing season; 3) what kinds of animals and how many.

A combination of a few AFA of surface rights to irrigate gardens, pastures, orchards, and a domestic residential water well with two AFA allowed use will meet the needs of most small family farm operations. You can irrigate with surface water, but you are going to want ground water for your domestic consumption, your culinary use.

For growth needs, you can present your future development plan for increased water consumption by irrigating more acres, adding orchards, pastures, gardens, greenhouses, or developing homesites with water needs. So long as you have a reasonable plan, extensions are generally granted. Deadline dates are very important.

CHAPTER 3

What Kind of Water? Irrigation Districts and Co-ops

WHAT KIND OF WATER DO YOU WANT AND NEED? The "manner of use" definition is dictated by the water designation by the state agency which controls all the water. If you are buying water rights you must know the designated use or manner of use allowed for that water. Will irrigation or agricultural water meet your needs? Do you plan to develop the land and sell lots with water? Will you be setting up for municipal or quasi-municipal use, or are you going to be using water for mining? These are not the same kind of water rights as those for farms and ranches. Farm water is often referred to as "agricultural water." The water agencies generally need to know where the point of diversion is located, what the water is being used for, exactly where is the place of use, how much water is being consumed, and time or season of use. You must file and receive state-level approval to change any of the water related facts. In the West, you cannot just divert water or pump water without the state's approval on these four issues (point of diversion, place of use, allocation amount, and manner of use). If it is seasonal irrigation water then time of use would be recorded. This can all be found on the state's water resources website. The public records will include current ownership, allocated amount, and allowed uses including a legal description of the place of use.

Ground water is the water pumped from the ground. Surface water is flowing on the surface in lakes, streams, ditches, etc. The state keeps records of both the ground and surface water. In an irrigation district, ownership may be shares of a private irrigation water system, a percentage of ownership in an organization which owns the water rights. These records at the small, private irrigation water districts can be more difficult to access and verify than your state's water records. We will come back to irrigation districts, water co-ops, and ditch companies.

In Western States, under the prior appropriation doctrine, priority dates are the "filed on" date, which is important under this system. The priority date is the date the water right was established by filing it. Many rights were established in the 1800's. There are still basins you can file for groundwater rights. This is known as "unappropriated water." Ask the state water resource agency if there are any of these basins available. You must to be prepared to establish this unappropriated water by putting it into beneficial use by pumping it. Generally domestic residential wells are allowed to pump up to two acre feet per year. That is approximately 650,000 gallons each year, a sufficient amount of water for many small family farms. Often the well driller will pull these permits from the state. The land owner is not required to seek additional permits for additional water rights on these domestic use residential wells. To confirm, check with your state engineer or division of water resources.

There are small agricultural water irrigation districts all over the Western United States. Most states can provide a list of these irrigation districts, and once you contact these water district managers or ditch companies, you'll find they are an excellent source of market data. They often know of members of the district who want to sell.

Typically members own water rights to surface water and they distribute it to the property owners who live in the district. This distribution system is most often a series of pipes, canals and ditches that convey water to the point of diversion on the canal, ditch, or stream where the owners divert it onto their property. This is all closely monitored by the owners and local water authorities. The owners along the ditch systems are allowed a set amount of diversion time each day, week, or month. If you are buying land in an irrigation district, you should understand the underlying water rights the district or water company owns in order to commit to delivery to landowners. There are many local names for this surface water, including ditch rights, run of the river rights, river rights, share rights, co-op water rights, etc. Many of the surface water rights are dependent upon the yearly cycles of sufficient rainfall and snowmelt on the watershed or drainage basin. Ask if the ditch has ever been low or dry.

Most of these water companies are non-profit corporations with bylaws, articles of incorporation, officers, directors, and membership, and they are subject to regular business meetings. The seller can provide all this information along with recent minutes and actions, budgets, liabilities, and assets. You need to know the following:

- Does the seller actually own the shares, and are they viable?

- Is the share you are buying allowed to be used on your land, when you plan to use it?

- Can the shares be conveyed clear of encumbrances like un-paid assessments?

- Will the certificate be properly endorsed to you and will it be re-issued in your name?

- What future maintenance costs and obligations to your new irrigation district might come up as an unpleasant surprise for you?

Considering all of this, you may want an attorney to review your share purchase into a water district.

CHAPTER 4

How Do You Find Water Rights?

PUBLIC RECORDS ARE AVAILABLE IN VIRTUALLY every state in which you can own water rights. Databases are not always the most user friendly and you may need to reference the specific state's glossary for the water division or department language and acronyms, but the basic principles primarily remain the same. The people within these agencies can be very helpful. When you start talking to them you should understand concepts like point of diversion, place of use, manner of use, and amount consumed.

- With irrigation water, time of use or season is important. **Point of diversion** may be your well or spring, or it might be where you block the channel in the river, stream, canal, or ditch system and force or divert the water onto your property.

- The **place of use** will be your area of consumption, garden, pasture, etc.

- The **manner of use** has to do with how you are using the water; are you mining, farming, gardening, etc.

- The **amount or allocation** is set forth in the water application, certificate, or share. Known as "duty," it is the amount measured by a meter on the well or by the simple means of the time allowed to divert your surface water.

Generally, any area can be researched by accessing these public records. The records can be extensive, but a good understanding of your water rights with the state is important. Filing new ownership records is required in some states and advised in every state. I have spent a great deal of time straightening out ownership records for my clients to be in a position to sell their water rights. Neither they nor their agent took the time to file properly on their water rights when they bought the land, and they may have owned them for thirty years. Make sure your new ownership is properly filed with the state agency and established in your name or you could lose your water rights.

I had a client who wanted to locate in a particular valley, but could not find the land they dreamed of for sale. We sat down and looked at the State Water Resources ownership pages for that basin. We found about thirty owners who had 100 acre feet or less of ground water rights in the area. All of this water was irrigation or agricultural water. This client wanted to irrigate a few acres for alfalfa hay for their horses. So we drafted a letter explaining briefly their interest in land in this area and sent it to all thirty owners. A gentleman who had received one of the letters called and explained that he had been thinking of selling his forty acres of pasture land with his old barn on it. He said he would be happy to meet the couple and discuss it with them. They made a fair deal, the farmer sold off his barn and some of his water, and the prospects found their dream land.

Sometimes you have to take the initiative to find what is not visible, yet still there right in front of you. This same principle goes for vacant land in an area you like; for land ownership records you go to the county assessor. Reach out, you will be surprised who may want to sell.

CHAPTER 5

Title Companies Do Not Insure Water

TITLE COMPANIES DO NOT INSURE WATER RIGHTS. Many counties do not yet recognize water in assessed value. There is not a breakdown of assigned value for water. However many water rights conveyances are done by the deed to the land. These records should be understood by the water buyer and their attorney or water engineer. These transactions can be found in the county records. Water is generally considered an appurtenance to the land, but a cautionary note, the land can be conveyed excluding all water rights. Water rights can be moved around, sold, traded, or it may have reverted back to the state for non-use. So check the public records. It is the state which keeps track of and regulates our water use at the highest level. The private irrigation districts referred to must also answer to the state level authority. Many banks will give value to water, or they will value irrigated ground higher than land with no water.

CHAPTER 6

How to Select a Land Sales Agent

FINDING THE RIGHT AGENT TO ASSIST YOU IN your search for land with water rights can be challenging. Most real estate agents have never listed or sold any water rights. However, there are agents who sell land in regions where the land is conveyed with water as an appurtenance, both surface and groundwater rights. The responsibility to get it right must fall back on you, the buyer. Caveat Emptor, or buyer beware, applies to water.

You are going to have to ask your agent more questions to determine if they are best able to help you. How many of their land sales have included water rights? Do they know which parcels listed for sale include water rights? Do they know how to include the water in the sale and then complete the conveyance with proper recordings with the correct officials and agency? This follow through is very important to you and your water right. These filings and recordings with your state water agency are going to be in addition to the normal real estate filings and recordings. The agent should be able to provide a list of the land offerings which include water rights and explain how much water and what kind of water is being offered. They should be able to show you past sales, including water, to use as comparable sales data. These past sales are used to establish a range of value for the water in your area. Water surveyors, the county assessor, and appraisers can be very helpful here. Your agent should know of these local professionals and be able to direct you to them. Do not assume that every agent is capable of helping. You are going to have to find the right agent. When buying, owners should provide all past correspondence with the water resource department, water application, certificate, place of use maps, etc. Also, you should know the priority date of your water rights. It establishes your place in the line of water rights, junior or senior. Remember: the older the better.

To protect your interests you should consider a water surveyor, engineer, or water attorney. Water abstracts or the chain of title for water can be complicated, and if it is complicated, then these professionals should be able to assist you. If you are buying a large parcel or multiple parcels with many water applications, permits, and certificates, you will need to consult a water engineer and water survey mapping service. They can help you with the conveyance or record of conveyance to be filed with the proper state agency.

CHAPTER 7

Who Keeps Track and What Else Matters?

WHO KEEPS TRACK AND WHAT ELSE REALLY MAT-ters? Here I would like to offer some examples, true stories about people and their water mishaps. I was working with local farmers in a designated basin, discussing the fact that their water table had dropped ten feet in the last two seasons. Upon further study, I found the following: the basin had allocations for municipal, agricultural, mining, and domestic or culinary rights totaling 150,000 acre feet per year. These are existing water applications, permits, and certificates issued by the state water authority. They are approvals to withdraw water from this basin. This is public record. Allocations are recorded and tracked by water resources in each state. Then with some diligence and digging I found a hydrologic study commissioned by the state in 1963 and completed in 1964. It stated that recharge to this basin from precipitation, rain and snow, combined with inflows from surrounding basins, the total annual recharge was estimated to be 30,000 acre feet. The basin recharges at 30,000 AFA and is being pumped or depleted at a rate of 150,000 AFA. It is over allocated by an estimated 120,000 AFA. After this evaluation the farmers no longer questioned why the water table had dropped. Now they asked just how much time they had left before they lost their farm and as a result their livelihood and lifestyle. You want to know this kind of information before you buy your land.

Another owner purchased land with a beautiful stream running through their forty acre parcel. They assumed they could divert some of the water for crops and animals. Shortly after they created a diversion point or dam and began using the water a state official showed up to explain that they had no rights to any water and must stop. They were given a cease and desist order. The water belongs to the state, he told them. Your rights, if you have any, are for the beneficial use and consumption of a set amount of water.

Yet another owner had a home on ten acres within an irrigation district. He had been on the property for more than ten years, watering his gardens, an orchard, and horse pasture land with his ditch water. This irrigation district actually owned a large irrigation well on an old farm down the road and piped the ground water to the canal system in the neighborhood. The well collapsed in the middle of irrigation season and his portion of the repair bill was over $12,000, due today, if he wanted water to his land. He did not understand his maintenance responsibilities and obligations associated with ownership of his water shares. Do your due diligence!

Land does not always have any water available, regardless of allocations or recharge rates. I watched an owner build a beautiful two-story log home on a five acre parcel, they planned to drill the well later. They finished the home at the same time the well driller drilled a 1000 foot dry hole. Then he drilled two more 1000 foot dry holes. After three years they had five dry holes on the property. Today, they still haul water to this home.

All these are common water stories and might have had different outcomes if they had checked the records and better understood the land, water, and their rights.

CHAPTER 8

Riparian or Prior Appropriation?

EAST OF THE 100TH MERIDIAN IN THE USA, when you buy land with water flowing past it, on and under it, it is yours to consume. These are riparian rights, sometimes known as "the rule of capture." Due to varying climate conditions in the United States two water right systems were established dictating the use of water known as prior-appropriation and riparian right. From Texas, north to the Dakotas, the 100th meridian is the major dividing line for rainfall in the United States. To the west of the line is arid to semi-arid states receiving very limited amounts of rainfall. In the Western States you must hold the rights in order to use the water. The system in effect here is known as "prior appropriation," a doctrine that operates on the simple principal of being in line. This applies to both the ground water we pump from a well, and the surface water we access and divert from a spring, stream, or pond. In the West you must have rights to use water, period!

Generally, East or West, riparian or prior appropriation, younger rights will not be allowed to impair older existing rights. This is "first in, last out." Since the states manage this valuable resource, the law, customs, systems, and rules vary from place to place. In some states today it is even illegal to collect the rain water from your roof.

Location

When you set out to buy water rights, you must first identify your specific region, basin, or designated area you plan to target, based on your state's mapping system. To understand the water balance in the target region, you need to know how much water has been allocated in this area. Next, you are going to research and find prior water studies, hydrology studies, and precipitation studies for this specific area. Most of our country has had water studies completed. In many areas you will find, as early as the 1940s, they knew plenty about water depletion. What you are looking for is estimated recharge rates for your area, basin, or region. Once you know how fast that imaginary bowl is being recharged or being naturally refilled, then you are going to compare this to the amount of current allocations in the area. Is it recharging as fast as it is being pumped out?

Is the table falling?

You are also going to do research into the water table studies for your area. Has it been stable or is there news of falling water table levels? If the basin is over allocated and the table has not yet started to fall, that does not mean it won't. Your research will involve you talking to the water well drillers and reviewing old well logs for your area. You may want to review Leonard F. Konikow's study, *USGS Groundwater Depletion in the United States (1900-2008)*. Well drilling logs for the many water wells drilled are available from the state and can provide a wealth of information. Hydrologists and geologists from the local university system can be very helpful, and often they are happy to talk to you about your area. You can talk with longtime residents about their well, the water table level, and how it is performing. One more example: in the spring one year, a couple located and bought a forty acre parcel of land with surface water rights. It had a small stream flowing across the land and diversion ponds. They built their home and farm. In the early fall, the water flow slowed to a trickle. Over the next five or so years more irrigation pivots and homes were built in the valley. It was growing in population, just like our planet. Today there is no water in their stream most of the year. With more careful evaluation they might have seen this coming, but they had no clue.

Basically you are going to understand your specific area, its water supply, and allocation. How much water is coming into the basin compared to how much is being removed? Is the basin balanced or over allocated and being depleted. You may be looking at the entire aquifer, watershed, or region for balance—regardless the same principles apply.

CHAPTER 9

First In, Last Out, Use It or Lose It and More

ALL WATER RIGHTS ARE NOT CREATED EQUAL, so the standing, or "durability," of your rights is important. Where do your rights stand in times of limited water access? Are they junior or senior? What are the limitations and guidelines issued by the state and the governing district for your rights? In the western part of America water right law dates back to mining and farming pioneers, and is based on the prior appropriation doctrine. In the East it is governed by "English Law" and the Riparian system. Many irrigation rights allow only seasonal use, so pay attention! There are sayings which generally do apply. "

- **First In, Last Out"** is prior appropriation—here priority dates always matter, and the older the better.

- **"Use it or lose it"** requires that most water rights are kept in beneficial use. The latter generally applies to the groundwater which must be pumped from a well, but not always; check with your state engineer.

- You may hear the term **"prove up"** water rights. This is essentially the time from application approval in which you are building improvements and putting the water to beneficial use to prove that you can use the amount applied for. This is important. It means that an approved application for 1000 AFA which only utilizes 400 AFA will not generally be approved as a permit or certificate for more than the 400 which has been kept in beneficial use.

When you come across unfamiliar terminology—the words may be related to hydrology, geology, engineering or state-specific jargon, or agricultural or any number of other industry terms—do not be intimidated; just ask what they mean.

Earl the water witch

In the late 1970s I was working fast and furious in the land development business and we were drillings lots of water wells. My favorite well driller was an old timer named Earl. He would show up on site with his rig and, first thing, he would pull out his pocket knife and go about finding the perfect branch. He had his own way of determining where he would drill. He would stand around carving his twig and listen to us and our advice about where we thought he should set up to drill. Then with two hands on his fresh branch he would start walking the property. When the branch began to bow down towards the ground, he would stop and say "this is the spot." He was water witching, or dowsing. If his fresh branch did not get results, he always carried his trusty brass rods. He had been doing this for forty years, and Earl knew more about water well drilling than anyone I had ever met.

I am not advocating water witching, although generally we took his advice and seldom missed our target. Rather, I want to share something Earl said to me back then that has stuck in my head ever since. He said "Chris, you pay attention to this water, because in your lifetime it is going to become more valuable than gold." I have met more than a handful of retired farmers and ranchers who have sold their farms, ranches, and most importantly their water rights, and every one of them now agrees with Earl!

How much should I pay?

Prices range from nearly nothing to lease a few AFA for a season, to tens of thousands per acre foot. Do your homework, understand the rights, the standing of the rights, the amount of water, the delivery system, and where you will use it. Find out what is for sale and what has sold, dig in and study this market. This is the local market price ranges in the specific basin or region.

If you use the outline presented here, you can learn the market and the values. The state water agency and county assessor can help you locate water, file on it in public records, transfer it, change water-related facts, and record your water ownership.

Ready to Start Looking?

You will be ready to start looking when you have an idea how you would like to put the water to beneficial use. You know if you are planning for residential consumption, animals, gardens, orchards, or pastures, if you plan to raise the food for animals in grass hay, grains, or alfalfa, or if your plans are for residential development or mining operations. You have an idea of how much water you will require. You are ready to find the correct kind of water (manner of use), in the location (place of use) you have chosen, and the allocation amount (duty) you will require.

CHAPTER 10

A River of Over Allocation

Millions of people in seven states and Mexico rely on the Colorado River for much more than their water supply. The Colorado River drainage basin or watershed is around 250,000 square miles, and the river runs 1450 miles. With considerations for environmental concerns, agriculture, industry, municipalities, and forty million people throughout Colorado, Utah, Arizona, California, New Mexico, Nevada, Wyoming, and Mexico all relying on the river for recreation, clean energy, food production, and water supply. Who gets to use the water, and for what purpose they consume the water, will continue to be hotly debated. The Colorado River is primarily fed by snowpack in the Rocky Mountains. The past few years have seen less snow and more dust than normal. This water is the lifeblood of the Southwest, and it is an example of a basin that is way out of balance in terms of allocation versus supply or recharge.

Agriculture, municipalities, environmental, and industrial needs are competing for this valuable resource. Each is important and each must have water. The Colorado River has become somewhat of the poster boy of water over allocation, conservation, distribution, and the effects of drought on our fresh water supply. Most people have heard about California's water issues, but you might be surprised to learn that many of our major principal aquifers across the country are in trouble. Due to the critical nature of water, there is no better time than now for you to secure your future water supply. As I've shown, there are very few absolutes in water—you must understand your area and check with your state authorities. The rules and regulations are in continual flux and change.

I sincerely hope this will help you secure your water future.

APPENDIX

Checklist

- Select region
- ASK: Prior appropriation or riparian?
- Obtain basin map and irrigation districts locations and contacts from Water Resources
- Narrow down your considerations to a basin or district, understand the watershed
- From Water Resources, locate total allocation for basin
- From local well logs and water managers, understand the water table or aquifer
- Consult with your professional contacts
- ASK: Do you understand the allocation versus recharge balance in the basin or watershed you are considering?
- From County Assessor or title company, locate property owners list
- From Water Resources, locate water rights holders

- Consider comparison pricing: irrigation District Managers, water engineers, land sales agents, attorneys, and Water Resources may know of relevant sales

- File or record your water rights purchase documents with County Recorder and Water Resources

ABOUT THE AUTHOR

Chris Miller has nearly 40 years of experience in the real estate industry, now specializing in land with water rights. His roots in the land go back generations, with a family that traveled the Oregon trail in a covered wagon to homestead on the rich soil of the Palouse region in southeastern Washington State. He has worked as a graduate assistant trainer for the Dale Carniegie Sales training program as well as an in-house training broker for a number of general realty companies and home builders. His work with land, resort developers, and home builders has taken him to New Mexico, Texas, Colorado, Wisconsin, Missouri, Arkansas, and Nevada. From condo conversions in Dallas, TX in the 70's to mountain retreats in Pagosa Springs, CO to private gated golf communities in Branson, MO, to new home and land sales for Arkansas' largest builder/land developer. In addition to real estate, Chris was educated as a financial advisor by Morgan Stanley Dean Witter. He writes extensively about real estate, finance, land, farm, ranch, agricultural lands, water rights, and land development.